WONDER STARTERS

Clocks

Pictures by TONY BARRS

Published by WONDER BOOKS
A Division of Grosset & Dunlap, Inc.

51 Madison Avenue New York, N.Y. 10010

Published in the United States by Wonder Books, a Division of Grosset & Dunlap, Inc.

ISBN: 0-448-09672-2 (Trade Edition)
ISBN: 0-448-06392-1 (Library Edition)

FIRST PRINTING 1973

Printed and bound in the United States.

Library of Congress Catalog Card Number: 73-1972

Our clock has stopped.
Daddy will fix it.

Most clocks have
lots of wheels inside them.
Daddy has taken all the wheels out.

Some clocks have glass sides.
You can see all the wheels inside.

Clocks tell people
what time it is.
The hands of the clock
point to the time.

Alarm clocks make a loud noise.
This tells people
it is time to get up.

This little clock
is called a watch.
People often wear watches
on their wrists.

This man carries
a watch in his pocket.
The watch has a chain.

Here is a very big clock.
It is in a church tower.
The clock has a bell
which chimes every hour.

There is a toy cuckoo
inside this clock.
It pops out every hour
and says "Cuckoo!"

This clock has big bells.
Every hour toy men
chime the bells.

The toys on this clock
dance every hour.

Here is an old
grandfather clock.
This clock tells the day
as well as the time.

This is a grandmother clock.
It is much smaller
than the grandfather clock.

Once there were no clocks.
Men told the time
by looking at the sun.

This is a sundial.
When the sun shines on it
there is a shadow.
The shadow points to the time.

Men used sand glasses
to tell the time.
The sand takes one hour
to run from top to bottom.

This is a water clock.
Water ran out of the clock slowly.
People could see how much water was left.
This told them the time.

This is a candle clock.
It takes one hour for the candle
to melt down to each mark.

There is a kind of clock
at this station.
It does not have hands.
It shows people what the time is.

Here is a clock
made of flowers.
There are different colors
for the hands and numbers.

This is a special watch
for blind people.
They can tell the time
by feeling the watch.

This is a clock shop.
People buy clocks here.
They bring broken clocks
to be repaired.

Starter's **Clocks** words

wheel
(page 2)

hand
(page 4)

alarm clock
(page 5)

watch
(page 6)

wrist
(page 6)

pocket
(page 7)

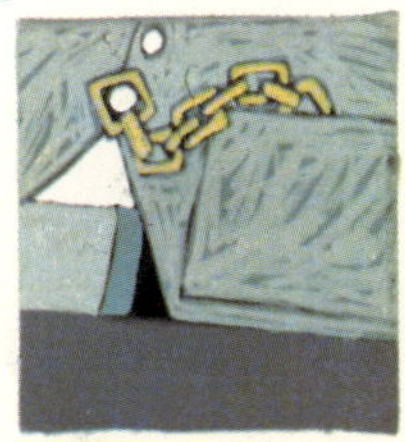

chain
(page 7)

church
(page 8)

tower
(page 8)

bell
(page 8)

cuckoo
(page 9)

toy
(page 10)

grandfather clock
(page 12)

grandmother clock
(page 13)

sun
(page 14)

sundial
(page 14)

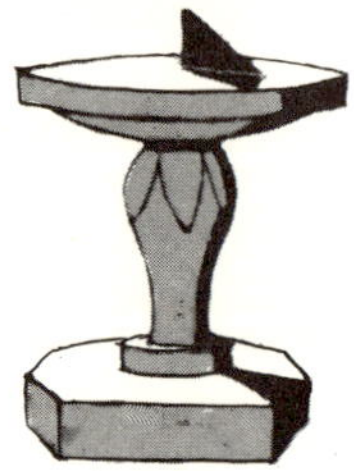

shadow
(page 15)

sand glass
(page 16)

sand
(page 16)

top
(page 16)

bottom
(page 16)

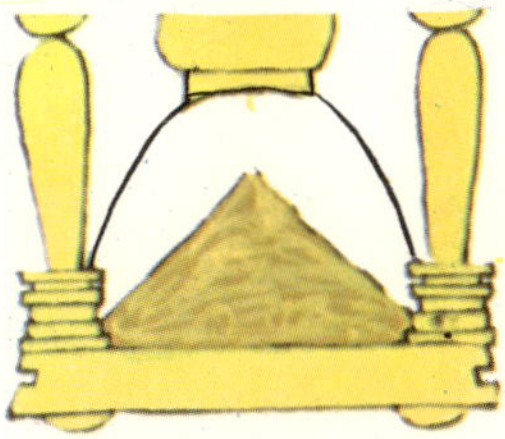

water
clock
(page 17)

water
(page 17)

candle
(page 18)

melt
(page 18)

station
(page 19)

flower
(pago 20)

numbers
(page 20)